MINECRAFT

MATHS
OFFICIAL WORKBOOK
AGES 6-7

D1493758

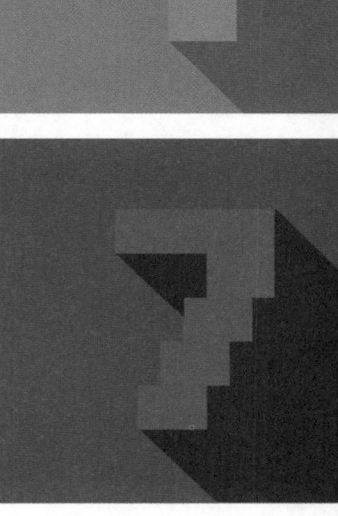

**DAN LIPSCOMBE
AND BRAD THOMPSON**

INTRODUCTION

HOW TO USE THIS BOOK

Welcome to an exciting educational experience! Your child will go on a series of adventures through the amazing world of Minecraft, improving their maths skills along the way. Matched to the National Curriculum for maths for ages 6–7 (Year 2), this workbook takes your child into fascinating landscapes where our heroes Jacob and Cali embark on building projects and daring treasure hunts...all while keeping those pesky mobs at bay!

As each adventure unfolds, your child will complete topic-based questions worth a certain number of emeralds. These can then be 'traded in' on the final page. The more challenging questions are marked with this icon to stretch your child's learning. Answers are included at the back of the book.

Note: While using this book, your child is likely to need some adult support, such as in reading explanations to them and giving any further help as necessary.

MEET OUR HEROES

Jacob loves making big things and small things – and everything in between! This busy lifestyle, constructing buildings and making tools, means he often works up an appetite. He really enjoys a piece of cake. His favourite colour is green...so finding emeralds on his adventures is just wonderful!

Cali adores exploring caves for precious ore. She rarely leaves home without her pickaxe. She is always wandering off when she spies an open cave and the shine of iron ore. Her favourite colour is gold...because of gold ore, of course! Cali also loves spending time with animals.

First published in 2021 by Collins
An imprint of HarperCollins*Publishers*
1 London Bridge Street, London, SE1 9GF

HarperCollins*Publishers*
1st Floor, Watermarque Building, Ringsend Road,
Dublin 4, Ireland

Publisher: Fiona McGlade
Authors: Dan Lipscombe and Brad Thompson
Project management: Richard Toms
Design: Ian Wrigley and Sarah Duxbury
Special thanks to Alex Wiltshire, Sherin Kwan and
Marie-Louise Bengtsson at Mojang and the team
at Farshore
Production: Karen Nulty

ISBN: 978-0-00-846275-8
British Library Cataloguing in Publication Data.
A CIP record of this book is available from the British Library.
2 3 4 5 6 7 8 9 10
Printed in the United Kingdom by Martins the Printers

MIX
Paper from responsible source
FSC www.fsc.org   FSC® C007454

This book is produced from independently certified FSC™ paper to ensure responsible forest management.

For more information visit: www.harpercollins.co.uk/green

CONTENTS

NUMBER, PLACE VALUE, ADDITION AND SUBTRACTION

TALL TREES IN THE TAIGA

The taiga is a chilly biome with tall spruce trees that grow close to each other. The trees have thick leaves circling their trunks. You may stumble across a curled-up, sleepy fox or even a wolf. You might be lucky enough to see a rabbit hopping around in the grass.

FOOD TO BE FOUND

The grass of the taiga is a darker green than in the plains. In patches, you will find ferns which can be cut down for wheat seeds. Pumpkins sprout up in groups and sweet berries grow on spiky bushes.

SEEING OUT THE NIGHT

Walking through the taiga at night can be dangerous. The tree leaves make it hard to see any mobs or holes in the ground. A tree can be chopped down to make a small hut to spend the night, but there is always a good chance of finding a village.

HOUSE HUNT

Today, Jacob is looking for a new place to call home. His last house was overrun with pillagers and he would prefer a quieter spot. As he walks the wilds of the Overworld, Jacob spots a group of buildings among the trees of the taiga. Perhaps he can set up a new house nearby.

ODD AND EVEN NUMBERS

It can be easy to get lost among the maze-like trees of the taiga. Fortunately, Jacob is good at finding his way and he continues in the right direction towards the village. He sees different types of animals.

Use this picture of the animals to answer questions 1–3.

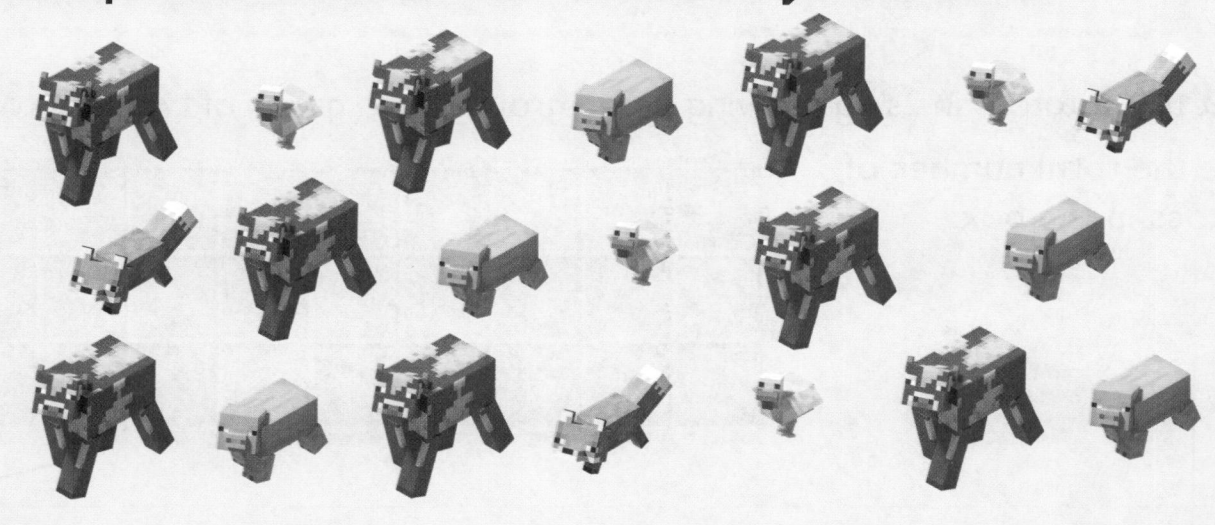

1

Count the cows.

How many cows can you see? Write the number in the box.

2

Jacob says that there are an even number of cows. Tick (✓) **true** or **false**.

True   False

3

Write **odd** or **even** for the number of pigs, chickens and foxes.

a)   Pigs   ....

b)   Chickens   ....

c)   Foxes   ....

COUNTING IN STEPS OF 2, 3, 5, AND 10

Jacob reaches the village. Some of the houses have small farm gardens. Most are growing potatoes and carrots. Others are growing wheat.

 1

Count the potatoes in 2s by drawing a ring around each group of two potatoes.

Write the total number of potatoes in the box.

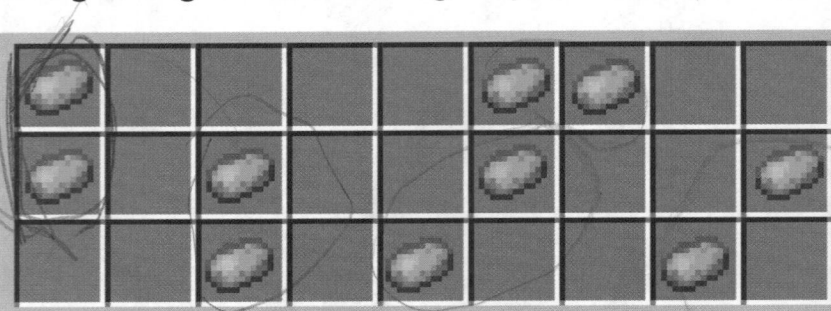

Once the wheat has grown, a villager can harvest it to make bread. For each loaf of bread, 3 wheat are used.

2

This picture shows how many wheat a villager is using to craft bread:

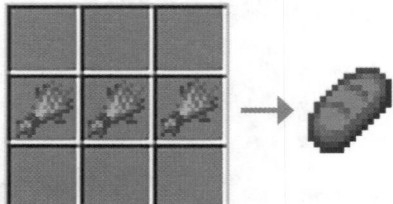 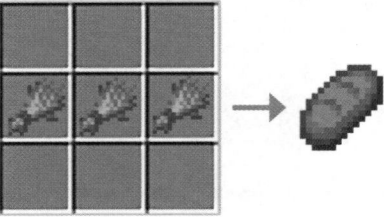 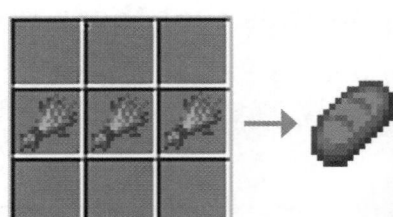

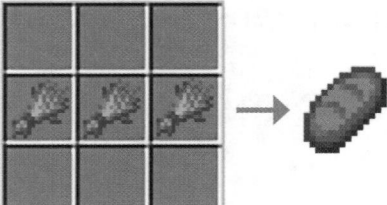 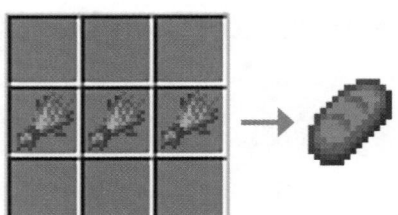

a) How many wheat are used in total?

b) How many bread can be crafted in total?

Jacob meets a villager who is dressed as a farmer. The villager has lots of carrots to sell.

3

Count the carrots in 5s.

How many carrots are there?

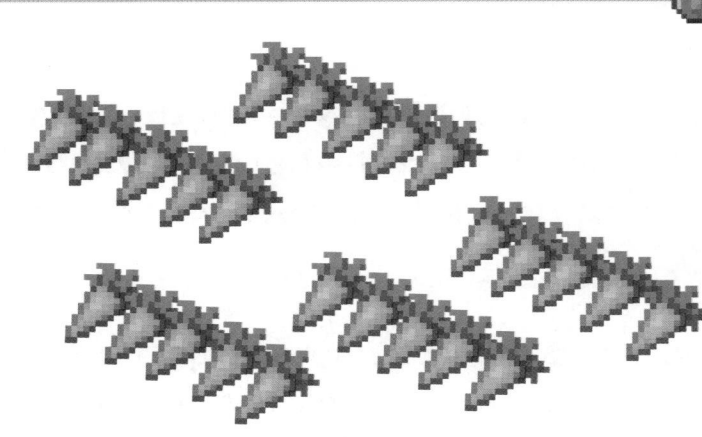

The villager has also grown pumpkins. They are in rows of 10.

4

 Count the pumpkins in 10s.

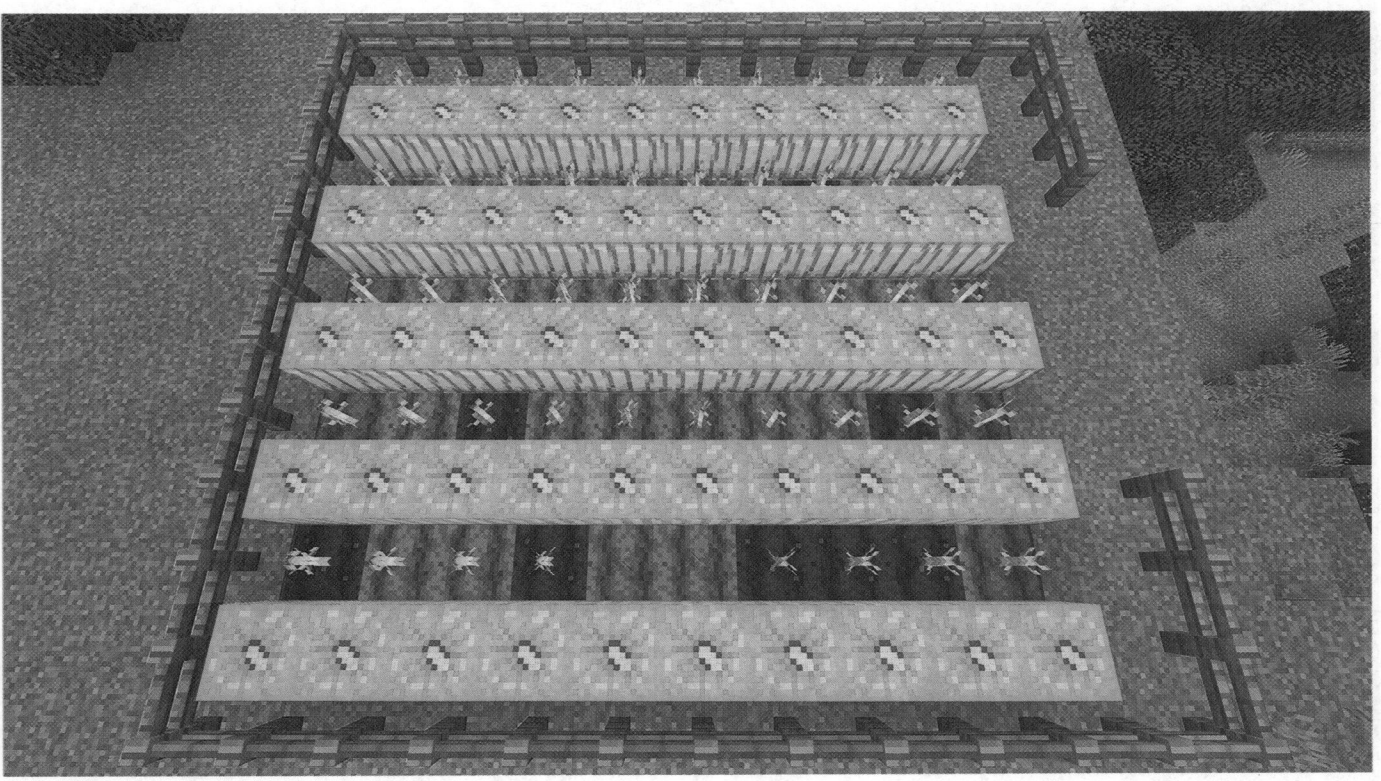

Write the total number of pumpkins in the box.

COUNTING FORWARDS AND BACKWARDS WITHIN 100

Jacob likes what he sees of the village. He wanders off to find a plot of flat land nearby where he will build a new house. Jacob starts building using the spruce wood around him.

The blocks in Jacob's house layout are numbered from 1 to 100. Five of the numbers are missing. Write these five numbers in the boxes below.

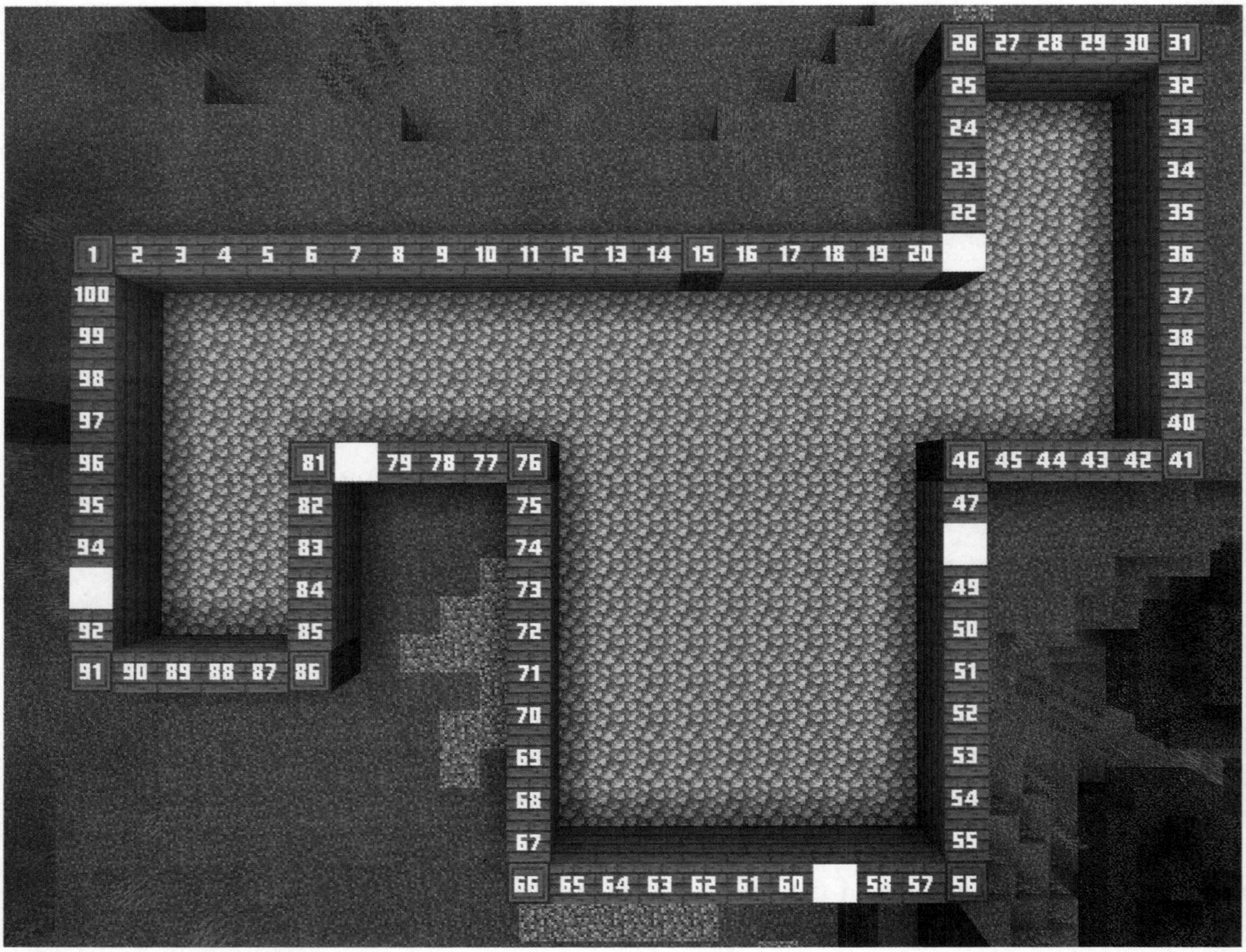

Jacob has built the lower level of his new house. He will soon add another floor on top. Before he does, he wants to mark out some of the land for a farm. He paces out some space, marking where he will build fences from the side of the house. Help him to do the counting.

 2

Jacob looks at block 67 on the house layout and counts forwards the given blocks.

Write the block he finishes on.

a) 7 blocks — 74

b) 13 blocks — 80

c) 19 blocks — 86

 3

Jacob looks at block 45 on the house layout and counts backwards the given blocks.

Write the block he finishes on.

a) 8 blocks

b) 15 blocks

c) 20 blocks

NUMBER REPRESENTATION

Jacob is thinking about his lunch break. He needs some coal to use as fuel in his furnace so he can cook meat. Finding a nearby cave, he takes his stone pickaxe to mine what he can.

1

What number is represented by each set of blocks? Draw lines to join the pictures to the correct numbers.

  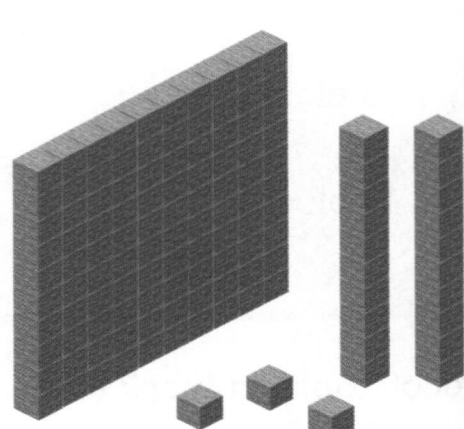

| 123 | 22 | 54 |

2

Write the numbers shown by the blocks.

a)

b)

 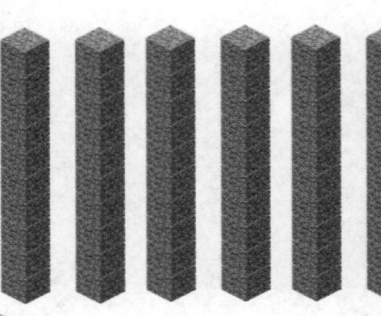

Jacob places some fuel and raw mutton into the furnace. While it cooks, he starts building the next floor of his house. The upstairs will have a bedroom, storage room and a library.

3

Jacob builds a section of his house using a total of 260 blocks.

Write how many hundreds, tens and ones there are in this number of blocks.

Hundreds [ ]   Tens [ ]   Ones [ ]

Jacob places his left-over blocks into storage chests. It's always worth keeping extra materials for future building. Jacob stores cobblestone, spruce planks and granite in a double chest.

4

Jacob stores 400 cobblestone, 80 spruce planks and 8 granite blocks in the double chest.

How many blocks does he store in total? Write how many hundreds, tens and ones.

Hundreds [ ]   Tens [ ]   Ones [ ]

COLOUR IN HOW MANY
EMERALDS YOU EARNED

LESS THAN, GREATER THAN AND EQUAL TO

Jacob places extra furnaces into the room he will use as a kitchen.

1

Tick (✓) the greater number of coal.

2

Look at the pictures. Write either **greater than** or **less than** in the space to make each sentence correct.

a)

The number of cows is .... the number of pigs.

b)

The number of chickens is .... the number of sheep.

Jacob uses bone meal to grow his crops quickly. He soon finds he has more than he needs. Jacob decides to trade some crops for emeralds.

3

The table shows how many of each crop can be traded for 1 emerald.

Jacob wants to trade: 20 wheat, 18 carrot and 30 potato.

Would 1 emerald be worth less than (<), greater than (>) or equal to (=) the value of Jacob's quantities of crops? Write **<**, **>** or **=** in each box to complete the statements.

Crop	How many of each crop can be traded for 1 emerald
Wheat	20
Carrot	22
Potato	26

a)

b)

c)

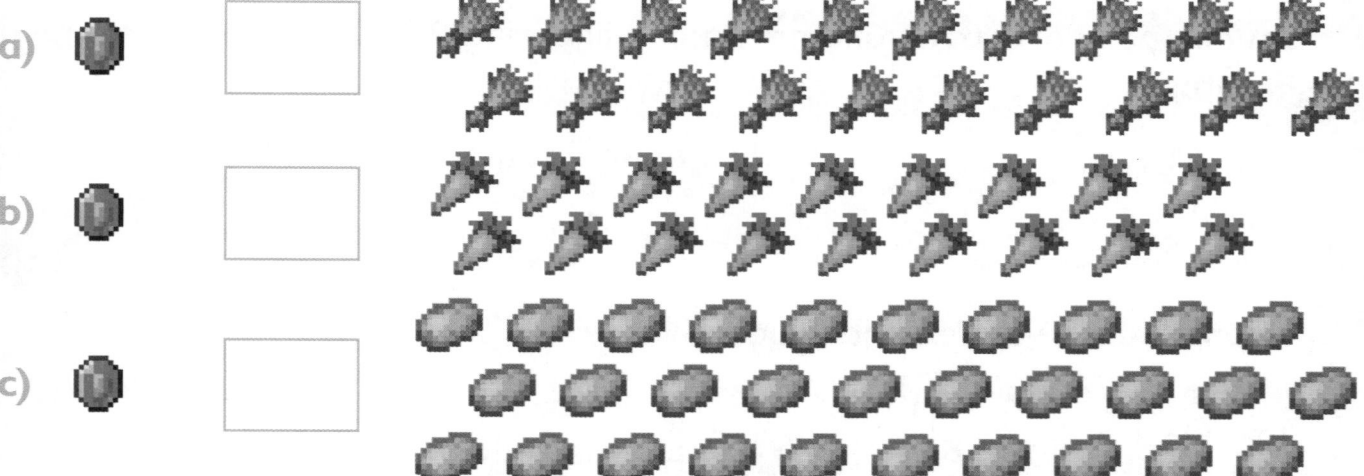

Jacob would really like some iron ingots to craft many other things.

4

The armourer says that 1 emerald can be traded for 4 iron ingots.

Would 2 emeralds be worth less than (<), greater than (>) or equal to (=) the value of 12 iron ingots?

Write **<**, **>** or **=** in the box to make the statement correct.

DOUBLING AND HALVING

Jacob hears a loud noise and the sound of arguing. Around the corner, he finds a group of pillagers fighting with some of the villagers. It looks like they are raiding the village. Luckily, Jacob has his weapons with him. First, he creeps over to the village bell and begins hammering it. This causes the villagers to run home. Now it's just Jacob and the pillagers.

I

Jacob crafted 8 poison-tipped arrows by using 8 normal arrows and I lingering Potion of Poison. Before he left home, he doubled this number of poison-tipped arrows.

How many normal arrows and how many lingering Potions of Poison did he use in total?

Normal arrows = ☐   Lingering Potions of Poison = ☐

2

Now complete the statements below.

a) Double I = ☐   I + I = ☐   I × 2 = ☐

b) Double 8 = ☐   8 + 8 = ☐   8 × 2 = ☐

Jacob approaches the pillagers with care as he prepares to attack. He is 24 blocks away but he wants to halve this distance.

3

Now complete the statements below.

Half of 24 = ☐   24 − ☐ = 12   24 ÷ 2 = ☐

Jacob fires his arrows and the pillagers fire back. While he is fighting, Jacob takes a lot of damage. Suddenly, he hears a familiar voice shout out. Cali has arrived to help!

4

a) Complete this number sentence to show how many arrows the two heroes fire.

$33 \times 2 =$

b) The fighting is over. Jacob and Cali won! The villagers reward them with a stack of 64 apples to share.

Complete this number sentence to show how the apples are shared.

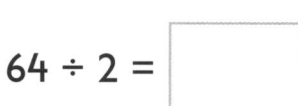

$64 \div 2 =$

After celebrating their victory, Jacob and Cali look for the pillager outpost. They find the tower and at the top of it is a chest. Inside the chest, Jacob finds iron ingots to add to his stock.

5

a) Complete this number sentence to show how many iron ingots Jacob now has:

$45 \times 2 =$

b) There was also some wheat in the chest. Jacob only needs half of the wheat. He gives the rest to Cali for helping him fight.

Complete this number sentence to show how many wheat Jacob now has:

$144 \div 2 =$

COLOUR IN HOW MANY EMERALDS YOU EARNED

SOLVING NUMBER PROBLEMS

Before Cali travels home, Jacob invites her to see his new house. He says she can help herself to some of his supplies for her next adventure. Cali needs to craft more arrows, so she takes some chicken feathers.

 1

a) Cali has 30 feathers in her inventory.

She takes another 34.

How many feathers does she have altogether?

b) Cali then uses 40 of the feathers to craft arrows.

How many feathers does she have left?

Jacob waves goodbye to Cali from his garden gate. He is safe behind his fences. He lays two polished granite paths in between his crops before going to bed.

 2

♥ **a)** Jacob has 78 polished granite.

He uses 36 polished granite for his first path.

How many polished granite does he have left?

b) Jacob then uses 38 more polished granite to make his second path.

How many polished granite did he use in total to make the two paths?

ADVENTURE ROUND-UP

AT HOME IN THE TAIGA

The taiga biome has become Jacob's home…for a while at least. That's the thing with adventurers — they often move to where the world needs them. For now, Jacob has a home with a kitchen, a cosy bedroom and all the storage he needs. He has made friends in the nearby village. He has also started growing crops.

ADVENTURES AHEAD

Jacob is always thinking about the next challenge. Talking to Cali made that feeling even stronger because she told him some of her exciting stories of adventure. Jacob will stay in the taiga a little longer. He will help the villagers to build their food stores. He starts to craft some armour and better weapons, ready for whatever the future holds.

A BOOK BEFORE BED

Jacob sits in his favourite spot in the garden. He has a torch next to him and his favourite book. Behind him, the sun sets, and the bats begin to flap around in the night sky. Jacob's eyes begin to slowly close; it is time for bed.

MULTIPLICATION, DIVISION AND FRACTIONS

OUT IN THE OCEAN

To explore the oceans of the Overworld, you must be incredibly brave…and prepared. To even see the full beauty and wonders, an adventurer needs to pack Potions of Water Breathing. Without these, you will run out of air quickly and pass out.

TREASURES… AND TROUBLE

As well as the danger of the water itself, drowned — a type of underwater zombie — will chase and throw tridents. Then there are guardians, large single-eyed creatures which haunt strange ocean monuments and will spike the unwary or shoot them with a beam of energy. There are wonderful materials everywhere. Prismarine sparkles and sea lanterns glow on the ocean floor.

WONDERS UNDER THE WAVES

Coral grows in brilliant colours, tropical fish swim among the usual cod and salmon. Feeding cod to a passing dolphin will encourage them to guide you to treasures held by sunken ships and monuments. Perhaps you will find a treasure map where X marks the spot of a priceless heart of the sea.

DIVING IN

Cali has worked up to this moment for a long time. She spent a long time catching pufferfish and using them to create a Potion of Water Breathing. Now she wants to swim with the tropical fish and explore the depths of the ocean.

MULTIPLICATION

The first thing Cali sees under the water is a colourful group of fish. This is called a shoal. Cali watches them dart around.

1

Cali counts the orange clownfish. Draw a ring around each group of two.

How many orange clownfish are there altogether?

6 groups of 2 = ☐

2

Cali counts the neon green fish. Draw a ring around each group of two.

How many green fish are there altogether?

5 groups of 2 = ☐

3

Cali finds some sea grass. It is tall and wavy. When she chops the sea grass with shears, it drops 2. Cali chops down 7 sea grass that each drop 2.

How much sea grass has Cali collected?

7 groups of 2 = ☐

4

Cali's helmet is enchanted with Aqua Affinity so she can mine more quickly underwater. She has found some chiselled sandstone. She collects 10 blocks, then another 10 blocks, and finally, another 10 blocks.

How many chiselled sandstone blocks has Cali mined altogether?

3 groups of 10 = ☐

COLOUR IN HOW MANY EMERALDS YOU EARNED

DIVISION

The coral is very pretty. Cali knows that it will die if she tries to collect it.

1

Cali can see 50 blocks of coral. Separate them into 10 columns of 5 by drawing them. The first set of 5 is shown.

Cali can see salmon, dolphins, cod and clownfish.

2

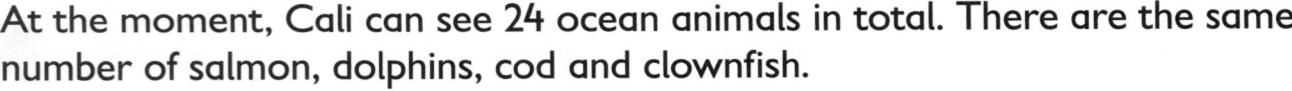

At the moment, Cali can see 24 ocean animals in total. There are the same number of salmon, dolphins, cod and clownfish.

Share out all 24 animals into equal groups by drawing them. The first animal in each group is already shown.

As Cali swims, she enters a shallow pool with patches of clay along the ground. She decides to stop and harvest some. Clay can be baked into bricks using a furnace.

3

Each clay block can be broken down into 4 clay balls.

Cali harvests 8 blocks of clay.

How many clay balls will Cali get from the 8 blocks?

Hopping out of the water, Cali grabs some sugar cane growing at the edge. Sugar cane can be broken down into sugar for baking. It can also be crafted into paper.

4

Cali has 30 sugar cane. She shares them out in groups of 3 to make paper.

Group the sugar cane into 3s by drawing rings around them.

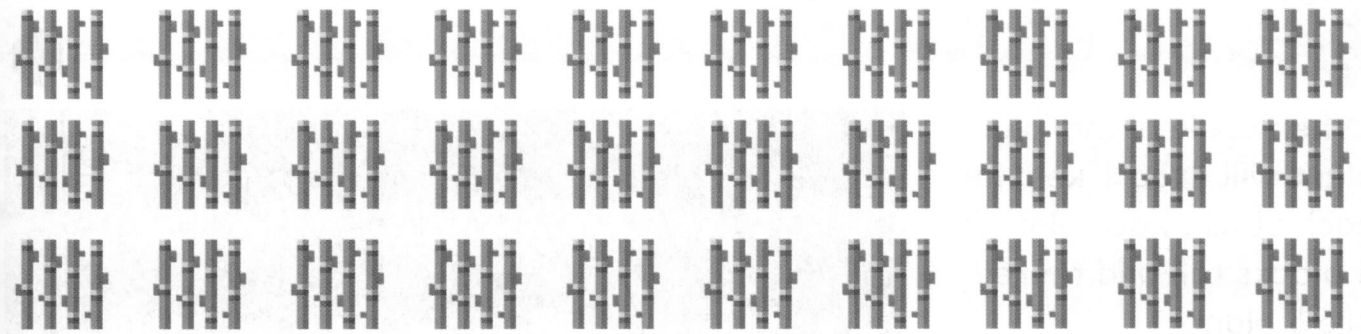

Complete the sentences to show how you have grouped the sugar cane.

30 grouped into 3s = ⬜ groups of 3

30 ÷ 3 = ⬜

2, 5 AND 10 TIMES TABLES

Cali swims back into the ocean, looking across the sea bed for treasures. A soft glow appears in the distance. Swimming down, she finds a large area covered with sponges and lit by sea lanterns.

1

Cali tries to harvest the sea lanterns, but they break into prismarine crystals. Each sea lantern breaks into 5 prismarine crystals.

How many crystals would appear if Cali broke 6 sea lanterns?

$6 \times 5 =$ ☐

Next, Cali spots some kelp. This plant only grows underwater. Cali would like to take some home. Kelp can be used as fuel in a furnace.

2

When broken, the kelp plant will drop 1 kelp for each block. Each plant is 5 blocks tall and there are 5 plants.

How many kelp does Cali have after harvesting?

$5 \times 5 =$ ☐

Cali swims past more coral. Her favourite colour of coral is purple. The coral stands 10 blocks tall.

3

Here are 7 columns of coral.

How many blocks of coral are there altogether?

$7 \times 10 =$ ☐

When she returns to the beach, Cali plans to mine 20 sandstone blocks from beneath the sand. She will use them to build pillars for her house.

4

Answer these questions to show how many pillars of different size she could make with 20 sandstone blocks.

a) Cali could craft ☐ pillars of 2 sandstone blocks.

b) Cali could craft ☐ pillars of 5 sandstone blocks.

c) Cali could craft ☐ pillars of 10 sandstone blocks.

DOUBLING AND HALVING

Cali feeds some cod to a passing dolphin. Once it has eaten, it zooms off leaving a trail of bubbles. Cali follows. She knows that this could lead to treasure. The dolphin stops above a sunken ship. The ship is broken in half and Cali can see two chests.

1

Both chests contain exactly the same items. Already shown below are the number of items in one of the chests. Double the number of items by drawing the same number of them in the space. Write the total in each box.

a)

b)

c)

2

So far, Cali has managed to avoid danger in the ocean. She has seen a lot of drowned, which seem to move around in pairs.

How many drowned does Cali see in:

a) 12 pairs of drowned?

b) 20 pairs of drowned?

c) 24 pairs of drowned?

Cali swims into another shallow pool and collects lily pads to lay in her pond at home. Help Cali work out how she will use them.

3

Cross out half of the lily pads in each set. Then complete the number sentence to show the half.

a)  ☐ ÷ 2 = ☐

b)  ☐ ÷ 2 = ☐

Cali builds a chest and places it at the water's edge to store her items.

4

♥ Cali is storing the clay she harvested. She wants to use the clay to craft some light blue terracotta. Cali can make 8 light blue terracotta blocks using 1 light blue dye.

Double 8 and then keep doubling your answer until Cali has 64 blocks of light blue terracotta.

8 → ☐ → ☐ → ☐

5

♥ Complete the sentences to show how many of each material Cali now has.

a) 9 blue dye are half of the
    total Cali has. She has a total of ☐.

b) 13 sandstone blocks are half of
    the total Cali has. She has a total of ☐.

c) 17 sugar cane are half of the
    total Cali has. She has a total of ☐.

SOLVING MULTIPLICATION AND DIVISION PROBLEMS

Cali dives back into the water with a splash! It's so relaxing being among the fishes.

1

Cali notices that other fish swim in groups of 3.

How many groups does Cali see if there are 6 fish?

2

Cali is surrounded by fish. After counting, Cali knows there are 48 fish. A dolphin darts among them and scares away half of the fish.

How many fish are left?

While Cali is distracted by the fish, some drowned move towards her.

3

To defeat a drowned, it must take 20 damage. Work out how many times Cali needs to hit the drowned with each weapon.

a) Wooden pickaxe:  = 2 damage → [ ] hits

b) Golden axe:  = 4 damage → [ ] hits

c) Wooden sword: = 5 damage → [ ] hits

d) Netherite sword: = 10 damage → [ ] hits

4

Look at your answers in question 3. Cali wants to defeat the drowned as quickly as possible. Which is the best weapon for her to use and why?

....

....

Cali notices that the drowned dropped a gold ingot. After defeating more drowned, Cali has 6 gold ingots to add to what she has at home. She will use them to craft gold boots.

5

Cali now has 24 gold ingots in total. She can make one pair of gold boots with 4 gold ingots.

Complete the number sentence to show how many pairs of gold boots she can make.

24 ÷ 4 = [ ]

WHAT IS A FRACTION?

In the deep water it can be hard to see clearly, but down below there seems to be a large building. As Cali swims closer, she is suddenly hurt!

1

A full health bar has 10 red hearts. Cali has half of her health bar left.

Tick (✓) the health bar that shows half.

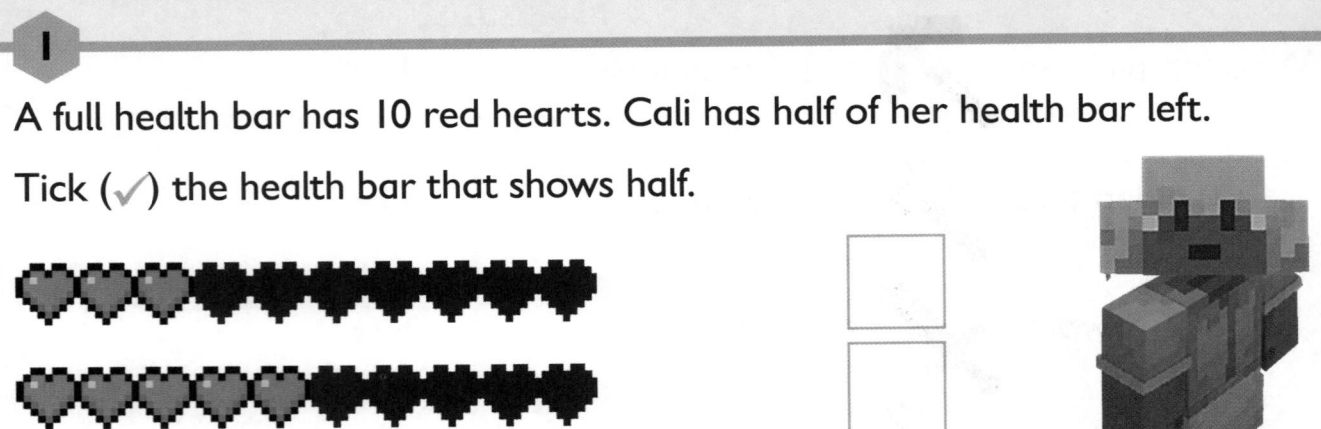

To one side, Cali notices a strange creature, which has one large eye and is covered in spikes. It's a guardian. It is firing its spikes at her! Cali has one quarter of her health left.

2

Shade one quarter of Cali's health bar.

3

A short time later, Cali only has 1 heart left in her health bar. To improve her health, she drinks a Potion of Healing. Each potion is worth 4 hearts.

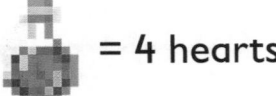 = 4 hearts

What fraction of the health bar does Cali have now she has had 1 Potion of Healing? Tick (✓) the correct answer. You can use the health bar below to help you.

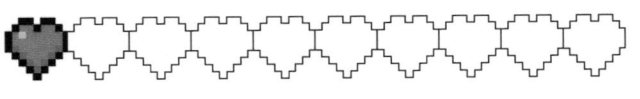

One quarter ☐   One half ☐   Three quarters ☐

Cali escapes from the guardian. Her health is back to full, but the swimming and fighting have made her hungry. Hunger is like health. I drumstick equals 2 points of hunger. So, 10 drumsticks equal 20 hunger points.

 4 ───────────────────────────────────────

Cali's hunger bar shows $2\frac{1}{2}$ drumsticks. So, she has 5 hunger points out of 20.

What fraction of her hunger bar is left? Shade the box with the correct answer.

| **Three quarters** | **One half** | **Two quarters** | **One quarter** |

With her tummy rumbling, some yummy cake would be perfect!

5 ───────────────────────────────────────

Here are 3 cakes. Tick (✓) the cake that has been split into thirds.

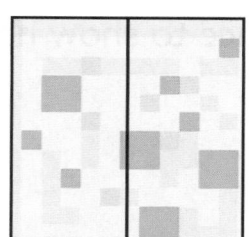 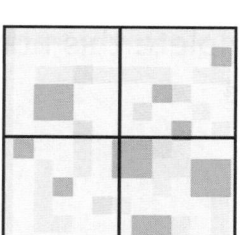 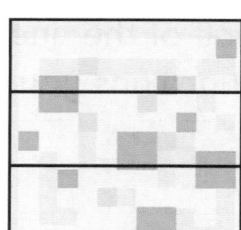

6

 Draw lines from each health bar to the fraction that shows how full it is.

| **One whole** |
| $\frac{3}{4}$ |
| $\frac{1}{4}$ |
| $\frac{1}{2}$ |

COLOUR IN HOW MANY EMERALDS YOU EARNED

FRACTIONS OF NUMBERS

Cali is curious about the guardians. They are dangerous and are protecting the large building. She swims a little closer. It's an ocean monument.

1

This picture shows the number of guardians around the monument.

a) Count the guardians around the monument.

There are ☐ guardians.

b) Circle half of the guardians. Complete this number sentence to show how many guardians you have circled.

$\frac{1}{2}$ of ☐ = ☐

Cali swims through the ocean monument, fighting guardians as they try to stop her exploring. She dodges laser attacks and sharp spines as they blast through the water!

2

The guardians all dropped prismarine shards. Cali has 20 prismarine shards now.

a) Circle one quarter of the prismarine shards and complete this number sentence.

$\frac{1}{4}$ of 20 = ☐

b) Circle three quarters of the prismarine shards and complete this number sentence.

$\frac{3}{4}$ of 20 = ☐

With fewer guardians around, Cali can explore the monument. She is running out of her Potion of Water Breathing, so she can't stay down here much longer. As she reaches the inner rooms, she is delighted with what she discovers.

3

Cali has found a treasure room! She sees 24 dark prismarine blocks. Write how many dark prismarine blocks would be represented by these fractions.

a) One quarter of 24 dark prismarine blocks is [ ] .

b) One half of 24 dark prismarine blocks is [ ] .

c) One third of 24 dark prismarine blocks is [ ] .

d) Three quarters of 24 dark prismarine blocks is [ ] .

Suddenly, a massive guardian swims into the room. This elder guardian is much stronger than the others.

4

The elder guardian is 32 blocks away.

Choose the correct answer to each question from the options in the box.

Halfway	One quarter	Three quarters

a) How close will Cali be if she swims 8 blocks closer?

....

b) How close will Cali be if she swims 16 blocks closer?

....

c) How close will Cali be if she swims 24 blocks closer?

....

COLOUR IN HOW MANY EMERALDS YOU EARNED

EQUIVALENT FRACTIONS

Cali would love to see more of the monument, but she is not strong enough to fight the elder guardian yet. She returns to shore and looks through her inventory.

 1

Here are two pictures of different inventories:

a) Cross out half of the items in the first inventory.

b) Cross out two quarters of the items in the second inventory.

Cali starts walking home. She has some cooked porkchops and apples left over from the start of the day. She decides to eat some and keep the rest, ready for tomorrow's exploring.

2

Cali has 12 apples and 12 cooked porkchops. She eats half of her apples and two quarters of her cooked porkchops.

Which does she eat more of? Show your working out.

....

....

....

COLOUR IN HOW MANY EMERALDS YOU EARNED

ADVENTURE ROUND-UP

OCEAN OUTING

Cali really enjoyed exploring the ocean. She saw parts of the world she had never seen before and lots of beautiful things. There were fish of all colours and shapes and sizes. They all swam around blocks of coral, which also shone brightly from sea lanterns and sea pickles.

DOLPHIN DELIGHT

It was brilliant to swim with the dolphins. One of them even guided Cali to a sunken ship! She would love to go back to the ocean soon and do more exploring. The monument looked amazing. She would love to search all the rooms inside, but she must get much stronger before she returns. That elder guardian was scary!

WATER FEATURES FOR HOME

Cali loves the water. She decides she will build more water features around her home. She can use lots of her new items for decoration. She has ideas swimming around in her head: a waterfall spilling into a prismarine pool, a pond full of lily pads and all of it bordered with tall sugar cane.

MEASUREMENT

EYEING UP AN ISLAND

Jacob has built a wooden boat and sailed across the ocean. On the horizon he sees some tall mushrooms on what appears to be a small island. He picks up speed to reach the shore of the mushroom fields.

MASSIVE MUSHROOMS

The mushroom fields biome is one of the rarest. It is usually found as an island. You can find purple-tinted ground known as mycelium and large, towering mushrooms rather than trees. The waters are also tinted, this time a light grey colour.

A STRANGE SORT OF COW

There is a rare animal only found on these types of island — the mooshroom. A mooshroom is a red and white spotted cow which has mushrooms sprouting from its back. Any adventurer who discovers these animals can breed them with wheat.

WEIRD AND WONDERFUL

An incredibly lucky hero may just climb ashore and discover a wrecked ship washed up on the beach. Exploring inside could reveal a treasure chest full of wonderful items. The mushroom fields are like something from a dream — hard to believe and delightful to explore.

LENGTH AND HEIGHT

As Jacob climbs from the water, the first thing he sees is the purple ground! That's not the only strange thing. Giant mushrooms are growing!

1

a) How many blocks high is this mycelium?

The mycelium is [ ] block(s) high.

b) 1 block is 1 metre high. How tall is this mycelium block?

The mycelium is [ ] m high.

2

The scale in the middle shows the height of each mushroom in blocks. Remember that 1 block is 1 metre in height.

a) How many blocks high is A? [ ]

How tall in metres? [ ] m

b) How many blocks high is B? [ ]

How tall in metres? [ ] m

A

B

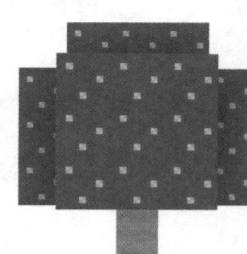

| 10 |
| 9 |
| 8 |
| 7 |
| 6 |
| 5 |
| 4 |
| 3 |
| 2 |
| 1 |

3

Use the ruler to measure each object.

a)

[ ] cm

b)

[ ] cm

COLOUR IN HOW MANY EMERALDS YOU EARNED

WEIGHT AND CAPACITY

Jacob begins chopping away at the tall mushrooms. With each swing of his axe, small mushrooms fall from the larger one. He will use them to craft some mushroom stew.

1

Look at the scales. Write the correct word from the box in each sentence.

heavier	lighter

a)

The tall mushroom is ....

than the small mushroom.

b)

The small mushroom is ....

than the mushroom stew.

2

Use the scales to find the weight of each object.

a)

☐ g

b)

☐ g

It's not long before Jacob finds a mooshroom. By milking the mooshroom with a bucket, he can get milk. By milking it with a bowl, he can get mushroom stew. It's thirsty work.

3

Look at these containers. It takes four water bottles to fill one cauldron.

 A  B  1  2

Choose the best word from the box to complete each sentence.

full empty twice half one two three four

a) Water bottle B is .... as full as water bottle A.

b) Water bottle A is ....

c) Cauldron 1 is ....

d) It would take .... full water bottles to fill cauldron 1.

e) It would take .... more water bottles to fill cauldron 2.

4

Each water bottle on the left is full and is going to be poured into the jug next to it. Shade each jug to show where the water will reach.

a)

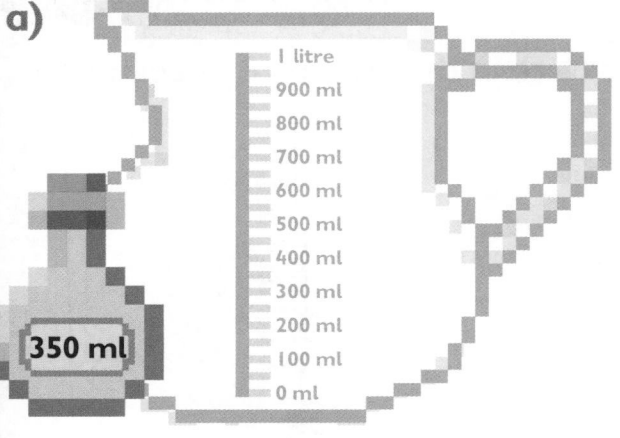

350 ml

b)

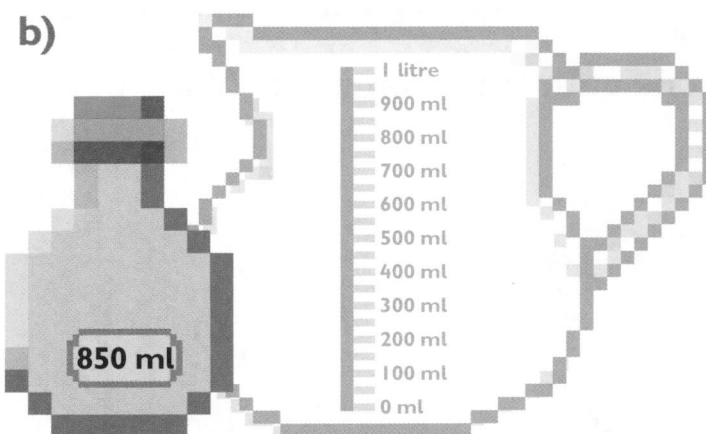

850 ml

TEMPERATURE

The mushroom fields biome is 'lush'. This means it is not too hot or too cold. It is a place where it can rain, but not snow. It's a comfortable temperature for Jacob.

1

Look at these thermometers and then complete the sentences.

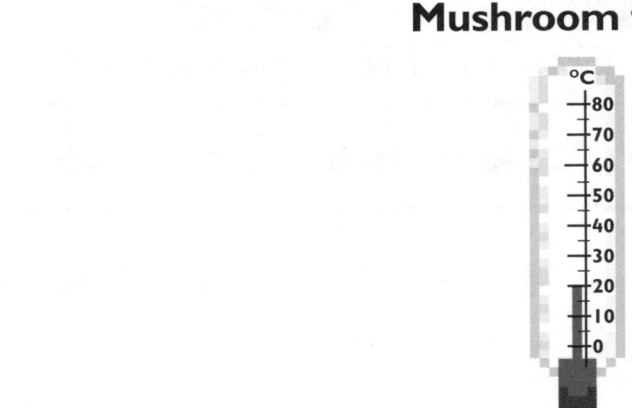

Desert

Mushroom fields

a) The temperature in the desert is [ ] °C.

b) The desert is .... than the mushroom fields.

c) The difference in temperature between the desert and the mushroom

fields is [ ] °C.

2

Look at the thermometers. Write the temperature in the box next to each one.

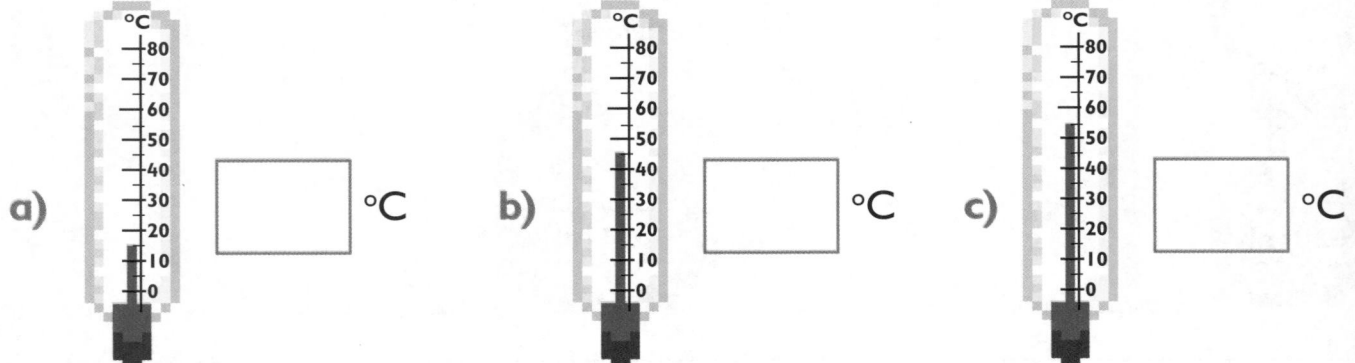

a) [ ] °C b) [ ] °C c) [ ] °C

The mushroom fields are one of the strangest places Jacob has ever seen. But soon it will be time to go home. The journey will take him through warm and cold conditions.

3

Colour each thermometer to show the temperatures.

a) 50°C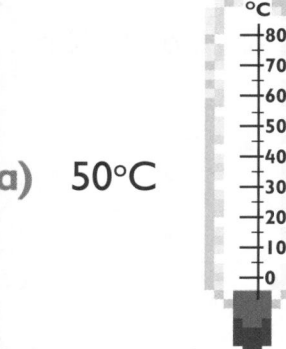

b) 35°C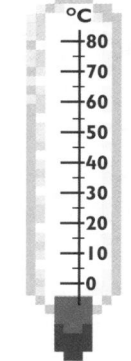

c) 65°C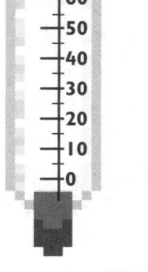

4

One night there was a temperature difference of 12°C between the mushroom fields and the desert.

What could the temperatures have been that night in the mushroom fields and in the desert? Find three pairs of possible answers.

[ ] °C and [ ] °C

[ ] °C and [ ] °C

[ ] °C and [ ] °C

COLOUR IN HOW MANY EMERALDS YOU EARNED

TIME

When Jacob checks his clock, it only shows sunrise and sunset. Can you show him how to tell the time on the clocks below?

1

Read the times shown on these clocks. Circle the correct time for each clock.

a)

12 o'clock   2 o'clock

b)

Quarter to 7   Quarter to 6

c)

Half-past 2   Half-past 5

d)

Quarter-past 3   Quarter-past 9

2

Using the clock, write how many minutes there are in the given time spells.

a) 1 hour = ☐ minutes

b) 2 hours = ☐ minutes

c) Half an hour = ☐ minutes

d) A quarter of an hour = ☐ minutes

e) Three-quarters of an hour = ☐ minutes

It's time Jacob made the journey home. Before he goes, he collects some more mushrooms. If only he could bring home a mooshroom too! While Jacob spends some time saying goodbye to the mooshrooms, take a few minutes to do these questions.

3

Write the missing numbers to show the minutes on a clock face. Each gap between the boxes is five minutes. Two have been done for you.

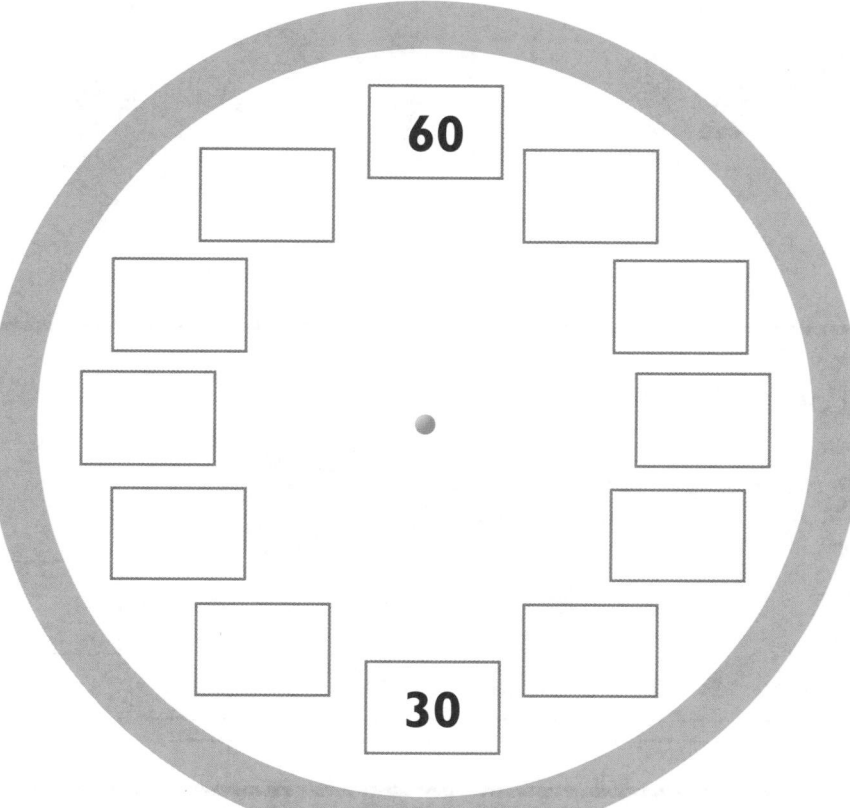

4

Look at the clocks. How much time has passed between the start and the finish?

Start   **Finish**

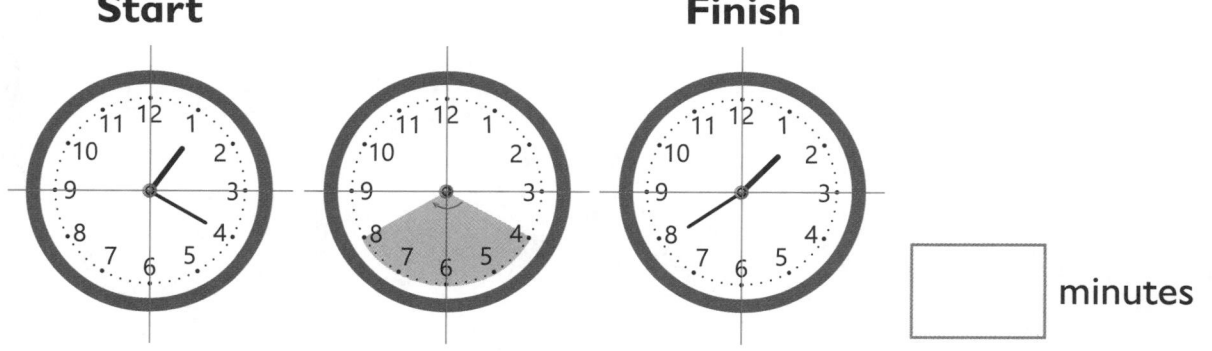

minutes

MONEY

Jacob stops at a village to trade some of the goodies he found in the mushroom fields. While he trades, show what you know about money.

 1

Write the value of each coin and note.

a)

b)

c)

....

2

Look at all the coins. Count the money.

£ [ ] and [ ] p

3

 Draw coins and notes to fill the piggy bank with a total of £17 and 55p.

COLOUR IN HOW MANY EMERALDS YOU EARNED

ADVENTURE ROUND-UP

A WELCOME FROM THE WOLF

Jacob eventually leaves the village and walks off to find his trusty wolf companion waiting at home. Before Jacob can even get through the door, his wolf jumps up to say hello. After a little fussing and some ear scratches, Jacob sits down to think about his trip to the mushroom fields.

FREAKY FIELDS!

Jacob did not expect to find such a wonderful and unusual place. The mushroom fields were very different from other biomes. With giant mushrooms and mooshroom cows, Jacob felt like he was dreaming! While he did not bring many items home, he did learn new things about his world, which is very important.

MOOSHROOM MEMORIES

If only Jacob could have brought home a mooshroom to live on his farm! What would the other cows have thought? The day had been peaceful – no fighting, no running away from mobs. Jacob had used a lot of energy though. Swimming and walking are great exercise but he's now tired. As he sits back and relaxes, his eyes slowly close, and soon he is fast asleep.

GEOMETRY AND STATISTICS

STEPPING INTO THE SHADOWS

Cali is out taking a gentle walk when she sees a large mushroom poking out from the trees ahead. She explores a little further and finds herself entering a dark forest.

FINDING A WAY THROUGH THE DARK

The dark forest is exactly what you might think. It is dark and full of trees. It is tough to walk through a dark forest. The tree trunks grow close together. There isn't any space for flowers to grow and they would never see sunlight anyway.

HIDING PLACES

There are a lot of mobs hiding among the leaves. Sometimes a zombie can be mistaken for a dark oak tree trunk. Or, as a hero wanders through, a creeper falls from the leaves above to stalk them. Meeting a creeper in the dark forest is especially horrid. There isn't much room to move when they explode, so it can be very dangerous.

LOOK ABOVE AND BELOW

Occasionally large mushrooms are found, like those in the mushroom fields. Caves await those who are not paying attention and swallow them up. However, once inside the caves, treasures may be found. Above ground there are riches too, but only the bravest of explorers will find them...

2-D SHAPES

Cali is a bit lost. There are so many trees packed closely together that they appear to create different shapes.

The shapes make Cali think that she could paint a nice picture of a forest scene to hang in her home. This is what she imagines:

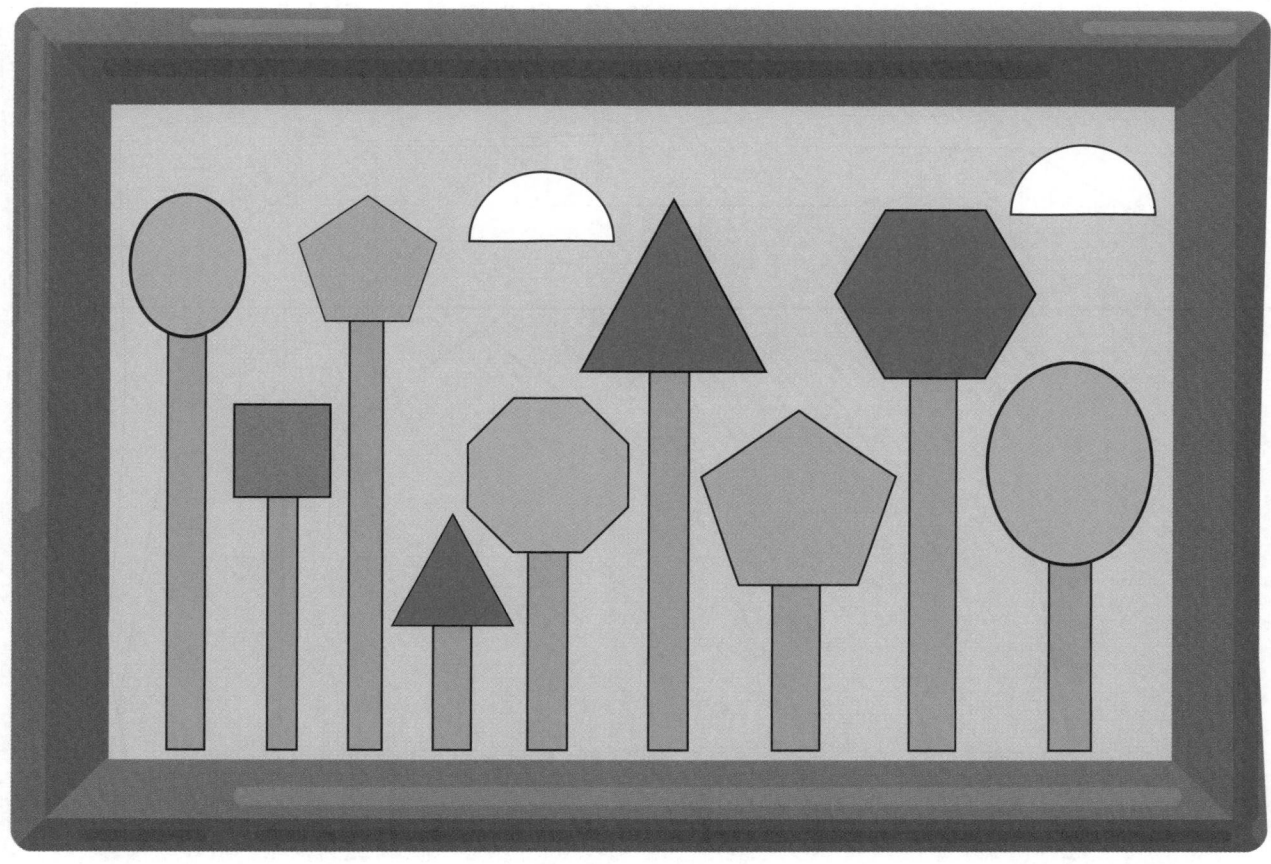

How many of each of these shapes can you see in the picture?

a) Squares ☐   b) Rectangles ☐

c) Triangles ☐   d) Semi-circles ☐

e) Ovals ☐   f) Hexagons ☐

g) Octagons ☐   h) Pentagons ☐

Cali wanders around for some time. She eventually reaches the end of the trees and a huge mansion comes into view. It's enormous! Cali is really interested in the mansion. While she makes her way to the front door, answer some more questions about 2-D shapes.

2

Below is a diagram of the mansion that Cali has found. Look closely at the shapes on the outside of the mansion. Some have a vertical line of symmetry – this means that if they are split in half down the middle, one half reflects the other.

Tick (✓) the six shapes that have a vertical line of symmetry.

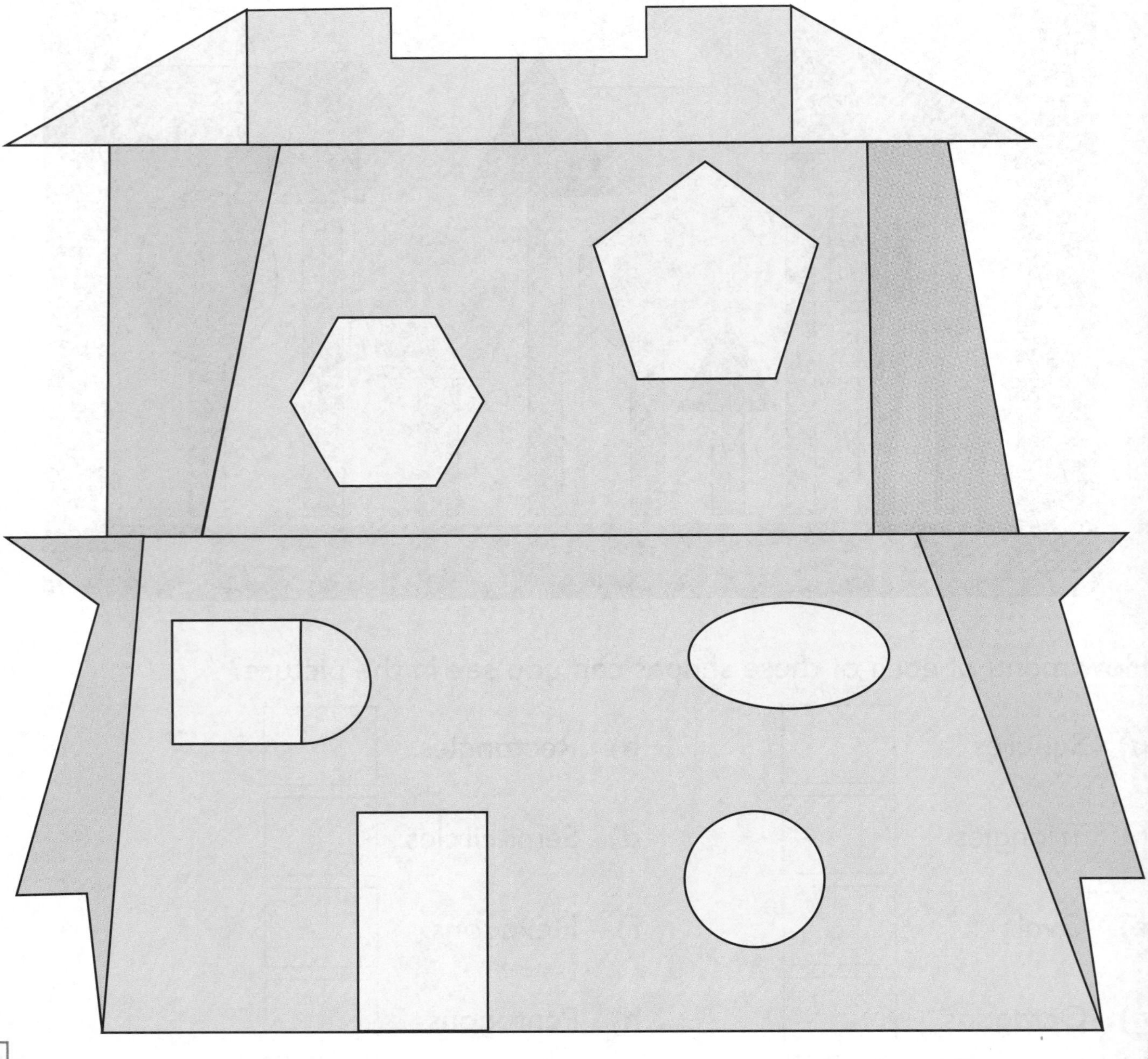

3

For each shape, fill in the table to show how many sides and corners it has. If the shape has a vertical line of symmetry, tick (✓) that column.

Shape	Sides	Corners	Vertical line of symmetry?

3-D SHAPES

Cali walks through the door and finds herself in a large room with several doors to the sides. Cali decides to explore. She picks a door and enters the room – it has a huge picture hanging on the wall.

1

The picture that Cali sees gives her an idea for making some sculptures. They would look something like these:

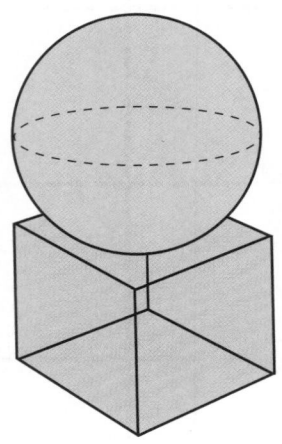
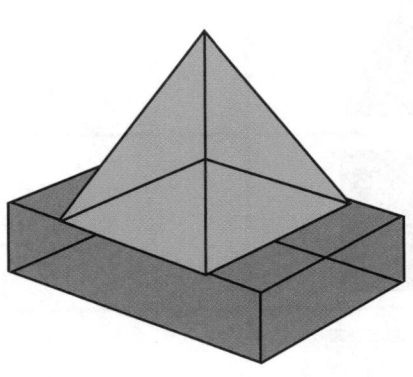

Tick (✓) the 3-D shapes that you can see in the diagrams of the sculptures above.

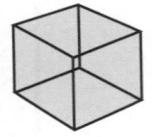  ☐   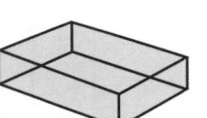 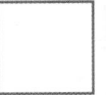 ☐   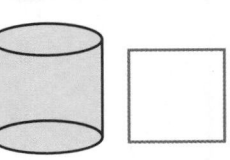 ☐   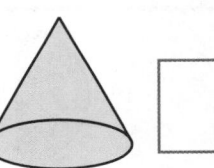 ☐

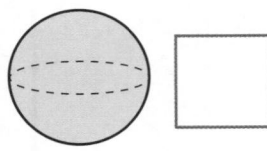 ☐   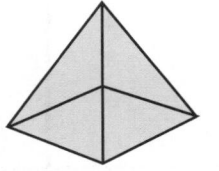  ☐    ☐

2

Write the name of the 3-D shapes shown here.

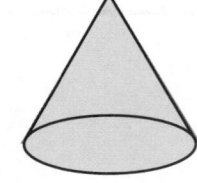

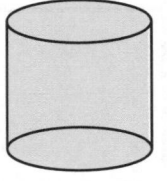

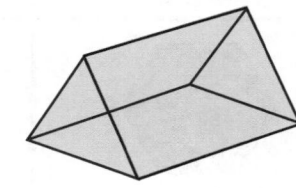

a)

b)

c)

....

The mansion is surprisingly quiet. Some rooms are almost empty but others contain weird and wonderful items of all kinds of shape. While Cali explores, move on to the next question.

3

❤ Complete the table to give the name and the number of faces, edges and vertices of each 3-D shape. Vertices are points or corners on a shape, such as where two edges meet.

Shape	Name	Faces	Edges	Vertices

PATTERNS

In one room, Cali admires the pattern made by a coloured floor. It makes the room feel cosy. Other rooms have patterns on the walls.

Look at the pattern of the floor in this room of the woodland mansion. Look at the different blocks used to build it.

Six blocks (shown in white) are missing from the pattern.

Colour in the blocks correctly to complete the pattern.

2

Look at the pattern on this wall in the woodland mansion. Look at the different blocks used to build it.

Six blocks (shown in white) are missing from the pattern.

Colour in the blocks correctly to complete the pattern.

After what seems a long time of opening doors and looking in rooms, Cali comes to an indoor garden full of melons, pumpkins and other crops. Someone must have planted them.

3

Look at the pattern on the roof of the woodland mansion. Look at the different blocks used to build it.

Six blocks (shown in white) are missing from the pattern.

Colour in the blocks correctly to complete the pattern.

The mansion also has a working farm. Different crops are growing.

4

Look at the pattern of the farm. Look at the different crops being grown.

Six blocks (shown in white) are missing from the pattern.

Draw and colour in the correct crops to complete the pattern.

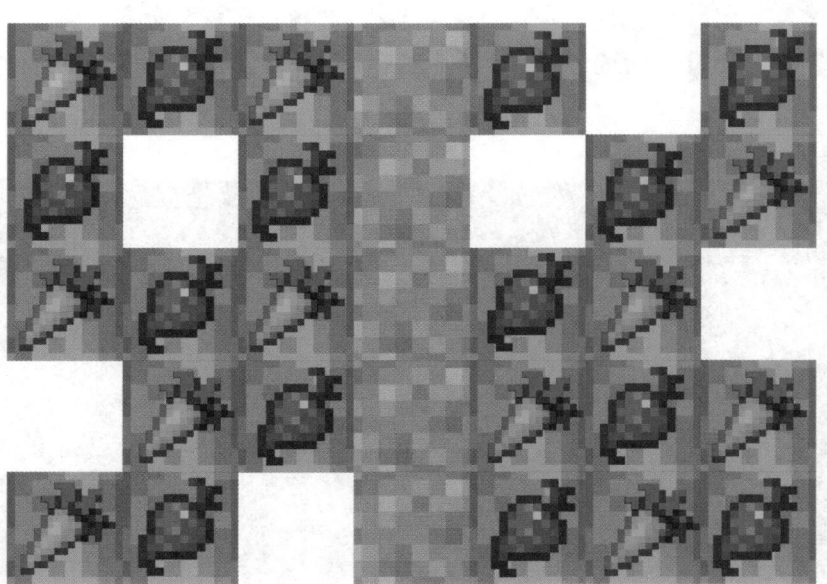

COLOUR IN HOW MANY EMERALDS YOU EARNED

SEQUENCES

Near the back wall, Cali finds places where mushrooms and flowers are being grown. She notices that they are planted in sequences. As she looks closer at them, somebody taps her on the shoulder…

 1

Look at the sequence of mushrooms and a fern plant. The sequence moves from left to right.

Draw and colour the next three.

   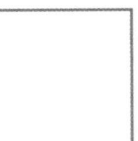 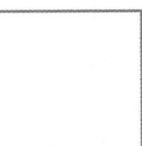 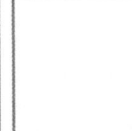

 **2**

Look at the sequence of flowers. The sequence moves from left to right.

Draw and colour the next three flowers.

     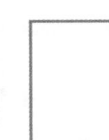  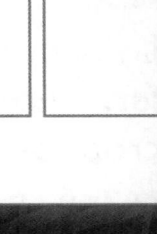

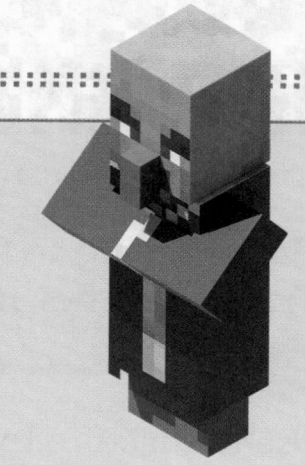

Cali turns to find an angry evoker! Suddenly, the floor erupts in a line of spikes. She begins to run around the evoker in a tight circle, swinging her sword as much as possible. The evoker falls and drops something shiny. It's time to escape. But this mansion is like a maze. Cali runs through the corridors, trying to find a way out.

3

The colours of the doors that Cali is running past look like they make a pattern. Or do they?

Is this pattern correct? How can you tell? ....

....

4

Draw a repeating sequence of your own on the grid below. The repeating sequence must contain no more than three different shapes.

FULL TURNS, QUARTER TURNS AND HALF TURNS

Cali is trying to find her way out of the mansion. It is all so confusing! Try to help her make the correct turns through the corridors.

1

Use this picture to answer questions 1–3.

				Bookshelf
				Jack o'lantern
				Flower pot
				Jukebox
				Torch

Cali is facing the flower pot.

a) Cali makes a full turn. She is now facing the ....

b) Cali makes a half turn. She is now facing the ....

c) Cali makes a quarter turn anti-clockwise.

She is now facing the ....

d) Cali makes a three-quarter turn clockwise.

She is now facing the ....

Some vindicators (illagers with iron axes) are giving chase. Help Cali to stay one step ahead of them by showing her the right directions. Will she escape?

2

Cali wants to move from the flower pot to the jack o'lantern.

Use a ruler to draw a possible route she could take on the grid on page 54. Assume that she can only move to a grid square that is directly above, below, to the left or to the right.

3

From her starting position shown in the diagram on page 54, Cali moves to the square that has the jukebox. From the jukebox she then moves to the square that has the bookshelf.

Use suitable words from the box to describe those movements.

| forwards   backwards   left   right   full turn   half turn |
| quarter turn   clockwise   anti-clockwise |

DATA AND INFORMATION

Cali finally finds a way out of the mansion and reaches safety. She looks at what she collected from the mansion. At first, Cali didn't think anyone lived there. If she had known, she wouldn't have taken anything.

1

The pictures show the number of different items that Cali managed to collect.

Use the pictures to record how many of each item she collected. Complete this tally chart:

Item	Tally
Book	
Slime ball	
Enchanted fishing rod	
Golden carrot	
Sticky piston	
Firework	

2

Complete the block graph to show the data shown in question 1. Shade in the boxes of the graph to show the number of items taken. Remember to check the scale.

8
6
4
2

3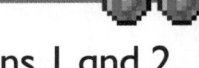

Answer these questions about the data you have recorded in questions 1 and 2.

a) How many more golden carrots did Cali take than books?

b) How many items did she take in total?

COLOUR IN HOW MANY EMERALDS YOU EARNED

ADVENTURE ROUND-UP

A CLOSE SHAVE

Cali finally leaves the forest behind. She is exhausted from running away from the illagers. If she had known they lived there, she wouldn't have gone near — at least not without preparing for a fight.

MYSTERY STATUE

Cali did manage to grab a number of items. At first, she feels bad for taking them, but she realises the illagers might have stolen them in the first place. She wonders if she can track down the owners of some of the items and give them back. An extra item that she found is a mysterious statue. When she gets home, she visits Jacob and asks him if he knows what it is.

A PRICELESS PIECE

Jacob takes one look at the statue and knows instantly what it is. It is a totem of undying. This is a very rare and magical item only found in woodland mansions. If the person holding it dies, it will bring them back to life…but only once. Both heroes look at the totem with shock and surprise. Cali will keep it safely in her inventory just in case she needs it during her next adventure…

ANSWERS

Page 5

1 8 [1 emerald]
2 True [1 emerald]
3 a) Odd [1 emerald]
b) Even [1 emerald]
c) Odd [1 emerald]

Pages 6–7

1 10 [1 emerald]
2 a) 15 [1 emerald]
b) 5 [1 emerald]
3 25 [1 emerald]
4 50 [1 emerald]

Pages 8–9

1 21; 48; 59; 80; 93 [1 emerald each]
2 a) 74 [1 emerald]
b) 80 [1 emerald]
c) 86 [1 emerald]
3 a) 37 [1 emerald]
b) 30 [1 emerald]
c) 25 [1 emerald]

Pages 10–11

1

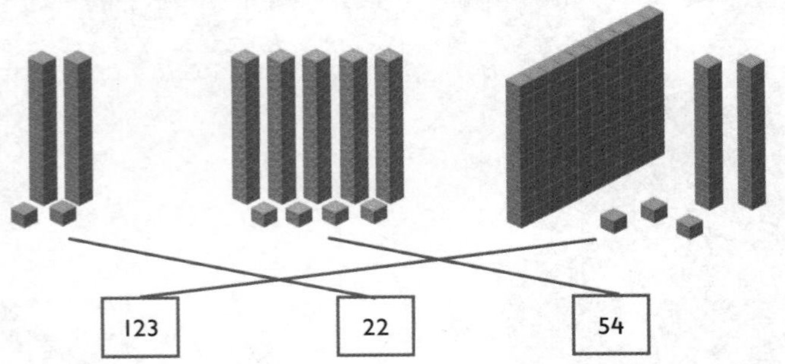

[1 emerald each]

2 a) 49 [1 emerald]
b) 261 [1 emerald]
3 Hundreds: 2
Tens: 6
Ones: 0 [1 emerald]
4 Hundreds: 4
Tens: 8
Ones: 8 [1 emerald]

Pages 12–13

1

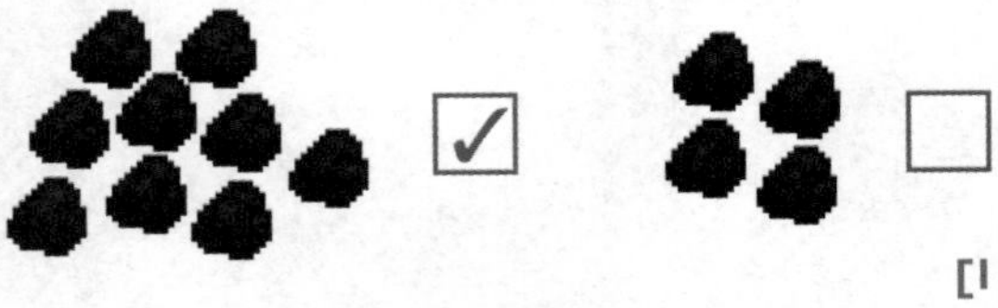

[1 emerald]

2 a) greater than [1 emeral
b) less than [1 emeral
3 a) = [1 emeral
b) > [1 emeral
c) < [1 emeral
4 < [1 emeral

Pages 14–15

1 Normal arrows = 16 [1 emeral
Lingering Potions of Poison = 2 [1 emeral
2 a) 2; 2; 2 [1 emeral
b) 16; 16; 16 [1 emeral
3 12; 12; 12 [1 emeral
4 a) 66 [1 emeral
b) 32 [1 emeral
5 a) 90 [1 emeral
b) 72 [1 emeral

Page 16

1 a) 64 [1 emera
b) 24 [1 emera
2 a) 42 [1 emera
b) 74 [1 emera

Page 19

1 12 [1 emera
2 10 [1 emera
3 14 [1 emera
4 30 [1 emera

Pages 20–21

1

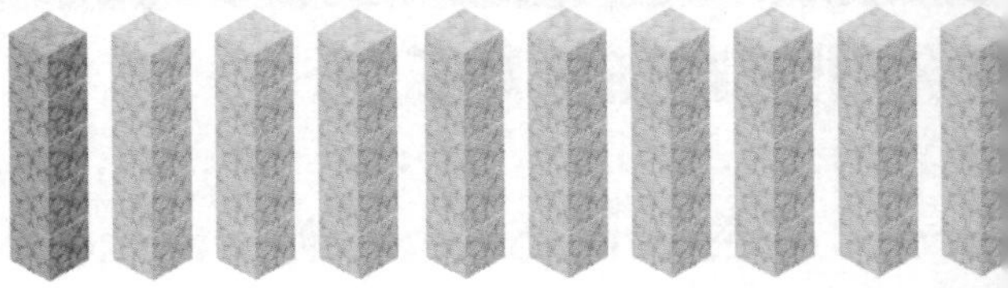

[1 emera

2

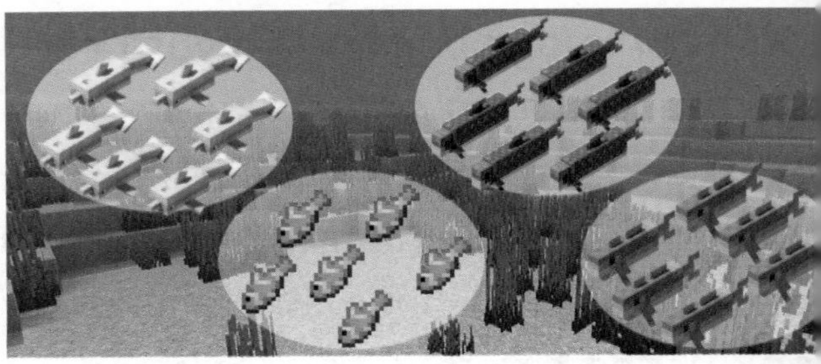

[1 emerc
3 32 [1 emerc
4 10; 10 [1 emerald ea

Pages 22–23

	30	[I emerald]
	25	[I emerald]
	70	[I emerald]
a)	10	[I emerald]
b)	4	[I emerald]
c)	2	[I emerald]

Pages 24–25

a)

 8 [I emerald]

b)

 10 [I emerald]

c)

 12 [I emerald]

a)	24	[I emerald]
b)	40	[I emerald]
c)	48	[I emerald]
a)	Any 6 lily pads crossed out. $12 \div 2 = 6$	[I emerald]
b)	Any 7 lily pads crossed out. $14 \div 2 = 7$	[I emerald]
	$8 \to 16 \to 32 \to 64$	[I emerald each]
a)	18	[I emerald]
b)	26	[I emerald]
c)	34	[I emerald]

Pages 26–27

	2	[I emerald]
	24	[I emerald]
a)	10	[I emerald]
b)	5	[I emerald]
c)	4	[I emerald]
d)	2	[I emerald]
	The netherite sword is the best weapon for her to use.	[I emerald]
	It will defeat the drowned in the fewest number of hits.	[I emerald]
	6	[I emerald]

Pages 28–29

  [I emerald]

Any $2\frac{1}{2}$ hearts should be shaded. [I emerald]
One half [I emerald]

4	Three quarters	[I emerald]

5 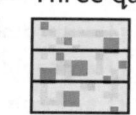 ✓ [I emerald]

6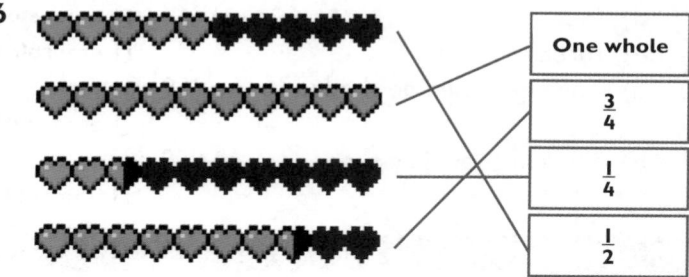

[I emerald each]

Pages 30–31

1	a)	16	[I emerald]
	b)	$\frac{1}{2}$ of $16 = 8$	[I emerald]
2	a)	5	[I emerald]
	b)	15	[I emerald]
3	a)	6	[I emerald]
	b)	12	[I emerald]
	c)	8	[I emerald]
	b)	18	[I emerald]
4	a)	One quarter	[I emerald]
	b)	Halfway	[I emerald]
	c)	Three quarters	[I emerald]

Page 32

1	a)	Any 8 items should be crossed out in the inventory.	[I emerald]
	b)	Any 8 items should be crossed out in the inventory.	[I emerald]
2		She eats the same number of apples and cooked porkchops.	[I emerald]
		$12 \div 2 = 6$ apples	
		One quarter of 12 cooked porkchops is 3, so two quarters is 6.	[I emerald]

Page 35

1	a)	1	[I emerald]
	b)	1 m	[I emerald]
2	a)	4; 4 m	[I emerald each]
	b)	6; 6 m	[I emerald each]
3	a)	6 cm	[I emerald]
	b)	8 cm	[I emerald]

Pages 36–37

1	**a)**	heavier	[1 emerald]
	b)	lighter	[1 emerald]
2	**a)**	450 g	[1 emerald]
	b)	800 g	[1 emerald]
3	**a)**	half	[1 emerald]
	b)	full	[1 emerald]
	c)	empty	[1 emerald]
	d)	four	[1 emerald]
	e)	three	[1 emerald]
4	**a)**	The jug should be shaded to the 350 ml mark	[1 emerald]
	b)	The jug should be shaded to the 850 ml mark	[1 emerald]

Pages 38–39

1	**a)**	40°C	[1 emerald]
	b)	warmer / hotter	[1 emerald]
	c)	20°C	[1 emerald]
2	**a)**	15°C	[1 emerald]
	b)	45°C	[1 emerald]
	c)	55°C	[1 emerald]

3 **a)** **b)** **c)**

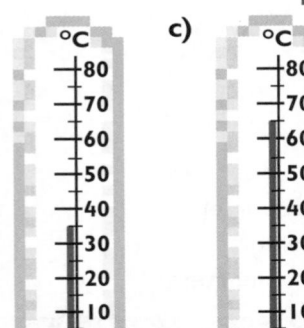

[1 emerald each]

4 Any three suitable answers.
Examples: 5°C and 17°C
10°C and 22°C
13°C and 25°C [1 emerald each]

Pages 40–41

1	**a)**	2 o'clock	[1 emerald]
	b)	Quarter to 7	[1 emerald]
	c)	Half-past 5	[1 emerald]
	d)	Quarter-past 9	[1 emerald]
2	**a)**	60 minutes	[1 emerald]
	b)	120 minutes	[1 emerald]
	c)	30 minutes	[1 emerald]
	d)	15 minutes	[1 emerald]
	e)	45 minutes	[1 emerald]

3

A clock face with numbers: 60, 55, 5, 50, 10, 45, 15, 40, 20, 35, 25, 30

[1 emerald eac

4 20 minutes [1 emera

Page 42

1	**a)**	£5	[1 emera
	b)	£2	[1 emera
	c)	£10	[1 emera
2		£3 and 99p	[1 emerald eac

3 Any combination of notes and coins to a value of £17.55

Example: one £10 note, one £5 note, one £2 coin, one 50p coin and one 5p coin

[1 emerald for £17 and 1 emerald for 55

Pages 45–47

1	**a)**	1	[1 emera
	b)	9	[1 emera
	c)	2	[1 emera
	d)	2	[1 emera
	e)	2	[1 emera
	f)	1	[1 emera
	g)	1	[1 emera
	h)	2	[1 emera

2

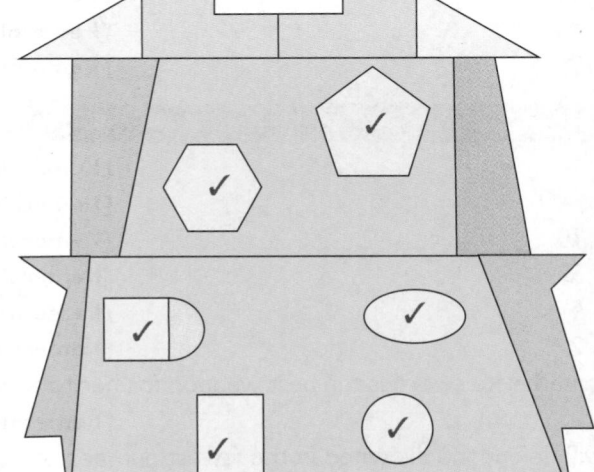

[1 emerald ea

Table completed as follows:

Shape	Sides	Corners	Vertical line of symmetry?	
circle	1	0	✓	[1 emerald]
triangle	3	3	✓	[1 emerald]
square	4	4	✓	[1 emerald]
rectangle	4	4	✓	[1 emerald]
semicircle	2	2	✗	[1 emerald]
oval	1	0	✓	[1 emerald]
pentagon	5	5	✓	[1 emerald]
hexagon	6	6	✓	[1 emerald]

ges 48–49

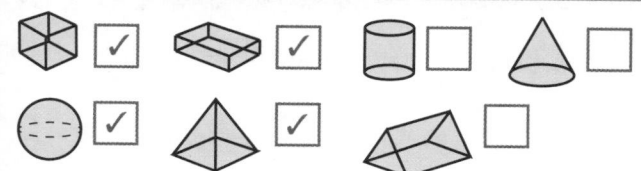

[1 emerald each]

a) Cone   [1 emerald]
b) Cylinder   [1 emerald]
c) Triangular prism   [1 emerald]

Table completed as follows:

Shape	Name	Faces	Edges	Vertices	
sphere	Sphere	1	0	0	[1 emerald]
cone	Cone	2	1	1	[1 emerald]
cylinder	Cylinder	3	2	0	[1 emerald]
cube	Cube	6	12	8	[1 emerald]
cuboid	Cuboid	6	12	8	[1 emerald]
pyramid	(Square based) pyramid	5	8	5	[1 emerald]
triangular prism	Triangular prism	5	9	6	[1 emerald]

Pages 50–51

1

[1 emerald for each colour correctly completed]

2

[1 emerald for each colour correctly completed]

3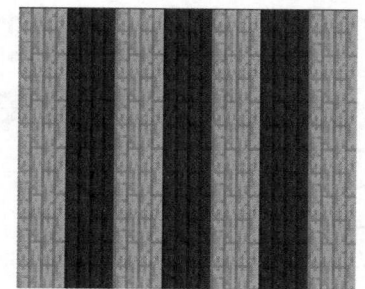

[1 emerald for each colour correctly completed]

4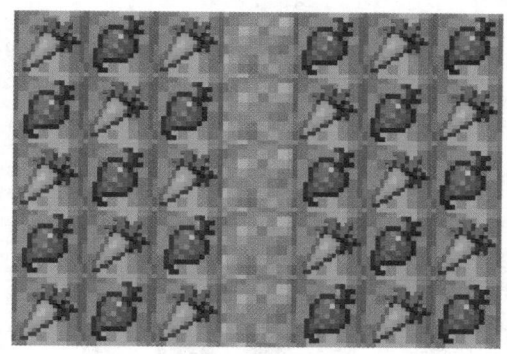

[1 emerald for each crop correctly completed]

Pages 52–53

1

[1 emerald each]

2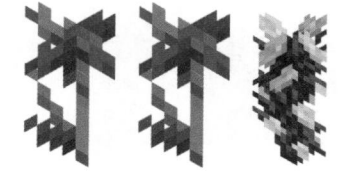

[1 emerald each]

3 The pattern is not correct.   [1 emerald]
   The last two doors on the right need to swap
   positions.   [1 emerald]

4 Any suitable repeating sequence that uses no more
   than three different shapes.   [1 emerald]

Pages 54–55

1 a) flower pot   [1 emerald]
   b) jukebox   [1 emerald]
   c) bookshelf   [1 emerald]
   d) bookshelf   [1 emerald]

2 Any suitable route drawn from the flower pot to the jack o'lantern which involves movement to squares that are directly above, below, to the left or to the right of each other.   [1 emerald]

3 Any suitable answer which takes Cali from her starting position to the jukebox and then to the bookshelf.

Example:

Cali makes a half turn. She walks forwards 3 squares. She then makes a quarter turn clockwise and walks forwards 3 squares. She then makes another quarter turn clockwise and walks forwards 3 squares.

[1 emerald for each correct movement up to a maximum of 5]

Page 56

1

Item	Tally
Book	III
Slime ball	IIII
Enchanted fishing rod	II
Golden carrot	IIII I
Sticky piston	I
Firework	III

[1 emerald for each correct tally]

2
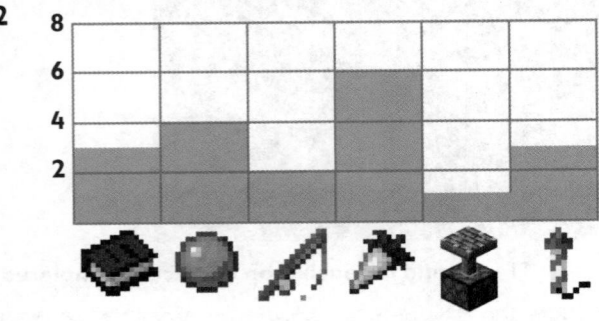

[1 emerald for each correct bar]

3 a) 3   [1 emerald]
   b) 19   [1 emerald]

TRADE IN YOUR EMERALDS!

Well done for helping Jacob and Cali to complete their adventures! Along the way, you earned emeralds for your hard work answering the questions. This villager merchant is waiting for you to spend your gems. Imagine you are setting off on an adventure of your own and trade your emeralds with the merchant for the things you'll need. What will you choose?

If you have enough emeralds, you could buy more than one of some items.

Ask a grown-up to help you count all your emeralds and write the total in this box.

HMMM?

SHOP INVENTORY

- IRON CHESTPLATE: 15 EMERALDS
- IRON LEGGINGS: 12 EMERALDS
- IRON HELMET: 8 EMERALDS
- IRON BOOTS: 6 EMERALDS
- DIAMOND CHESTPLATE: 30 EMERALDS
- DIAMOND LEGGINGS: 24 EMERALDS
- DIAMOND HELMET: 16 EMERALDS
- DIAMOND BOOTS: 12 EMERALDS
- SHIELD: 20 EMERALDS
- BELL: 5 EMERALDS
- ENCHANTED IRON CHESTPLATE: 25 EMERALDS
- ENCHANTED IRON BOOTS: 10 EMERALDS
- ENCHANTED DIAMOND BOOTS: 20 EMERALDS
- ENCHANTED DIAMOND CHESTPLATE: 50 EMERR
- ENCHANTED DIAMOND LEGGINGS: 40 EMERALD

That's a lot of emeralds. Well done! Remember, just like real money, you don't need to spend it all. Sometimes it's good to save up.